I0470220

IF FOUND PLEASE RETURN TO

....

....

....

EMAIL ....

PHONE ....

CONTENTS • SUBJECT	PAGE

Christian Holthaus • Juifenstr. 1 • 81373 München
Printed by
Amazon Media EU S.à r.l., 5 Rue Plaetis, L-2338, Luxembourg

www.ingramcontent.com/pod-product-compliance
Lightning Source LLC
Chambersburg PA
CBHW020921180526
45163CB00007B/2827